THE BOOK OF ANSWERS

神秘的

答案之书

韦恩 · 著

江苏凤凰文艺出版社
JIANGSU PHOENIX LITERATURE AND ART PUBLISHING, LTD

本书使用说明

1. 把书放在桌上，闭上眼睛。

2. 用5至10秒钟的时间集中思考你的问题。例如："他喜欢我吗？"或"我需要换个工作吗？"

3. 在想象或说出你的问题的同时（每次只能有一个问题），一只手放在这本书的封面上，从后往前，翻动书页的边缘。

4. 在你感觉时间正确的时候，打开书，你要寻找的答案就在那里。

5. 遇到任何问题，你都可以翻开它。

打开它

找到你的人生解答。

勇 于 承 担
将 会 获 得 好 的 结 果

A strong commitment will achieve good results

谨 慎 处 理

Approach cautiously

A

绝对地

Definitely

A

履行你应尽的义务

Follow through on your obligations

如果你不抵触的话

If you don't resist

A

这并不值得
放手一搏

It's not worth a struggle

A

认真倾听
你就会知道

Listen more carefully
then you will know

A

无论怎样

No matter what

A

如果你能说“谢谢”

Provided you say “Thank you”

A

灵 活 变 通

Remain flexible

A

意料之外

Startling events

A

这是你
永远不会忘记的

It is something you won't forget

A

那将仍是不可预测的

It will remain unpredictable

很有可能发生意外

Mishaps are highly probable

A

其他事物
将取决于你的选择

Others will depend on your choices

重新斟酌你的方法

Reconsider your approach

A

与另一种状况 息息相关

There is a substantial link to another situation

A

相信你的直觉

Trust your intuition

走 路 要 小 心

Watch your step as you go

A

你一定会得到支持

You are sure to have support

你的行为
会使事物更加美好

Your actions will improve things

局势尚未清晰

The situation is unclear

A

大声说出来

Speak up about it

A

找 出 更 多 办 法 来

Seek out more options

立即停止

Press for closure

A

行动起来

Move on

A

笑一笑

Laugh about it

A

貌似是有把握的

It seems assured

A

调查了解之后
投入其中

Investigate and then enjoy it

不 要 浪 费 你 的 时 间

Don't waste your time

别去赌

Don't bet on it

A

耐心一点

Be patient

A

绝对不

Absolutely not

A

大方一点

Be more generous

A

不要被压力
迫使着行事

Don't be pressured
into acting too quickly

A

先去完成其他事

Finish something else first

A

付 出 会 有 回 报 的

Gentle persistence will pay off

A

那将是意外的

It could be extraordinary

A

以好奇之心去探究

Explore it with playful curiosity

A

那将是一件乐事

It will be a pleasure

A

别犹豫

Don't hesitate

A

此 事 不 可 取

It would be inadvisable

拟一个清单
说明为什么

Make a list of why

A

一 人 难 成 事

Not if you're alone

A

重 新 考 量 一 下
什 么 才 是 重 要 的

Reprioritize what is important

三思而后行

Take more time to decide

A

答 案 就 在 你 身 后

The answer is in your backyard

A

你需要考虑
其他办法

You'll need to consider other ways

A

有乐观的理由

There is good reason
to be optimistic

你将得到支持

It will sustain you

A

并非最佳时机

Unfavorable at this time

A

不要被情绪左右

Don't get caught up
in your emotions

A

拭目以待

Watch and see what happens

A

你会发现
自己无法妥协

You could find yourself unable to compromise

A

带着你的善意
坚持到底

Follow through with your good intentions

尝试剑走偏锋

Try a more unlikely solution

毋庸置疑

Unquestionable

A

为确保做出
最好的决定，
请冷静

To ensure the best decision, be calm

A

其 实 你 不 是 很 在 意

You don't really care

相信你最初的想法

Trust your original thought

A

发挥你的想象力

Use your imagination

A

等一个更好的答复

Wait for a better offer

A

等 一 等

Wait

A

需要付出
足够的努力

*A substantial effort
will be required*

A

你将不得不
放弃其他事物

You may have to drop other things

A

对此表示怀疑

Doubt it

A

赌 一 下

Bet on it

A

一年后这件事
就没那么重要了

A year from now it won't matter

A

跟随他人的引导

Follow someone else's lead

A

把它写下来

Get it in writing

A

记 得 找 点 乐 子

Don't forget to have fun

接受一个
日常上的改变

Accept a change to your routine

A

你需要更多信息

You'll need more information

A

鼓起勇气

Adopt an adventurous attitude

A

此时别再索求

Don't ask for any more at this time

A

把精力放在
家庭生活上

Focus on your home life

A

让自己先休息一下

Allow yourself to rest first

A

最好再等等

Better to wait

A

合 作 将 是 关 键

Collaboration will be the key

你必须立即行动

You must act now

期 望 解 决

Expect to settle

A

欣喜地确认这一点

Be delightfully sure of it

别傻了

Don’t be ridiculous

A

表现得就好像这已经是真的了

Act as though it is already real

A

别怀疑

Don't doubt it

A

问 问 你 爸

Ask your father

A

听 听 过 来 人 的 意 见

Follow the advice of experts

辅助将会让你的项目获得成功

Assistance would make your progress a success

你将会失望

You will be disappointed

A

尽早去做

Do it early

A

问问你妈

Ask your mother

A

规 避 第 一 种 方 案

Avoid the first solution

A

实际一点

Be practical

A

把这当作一次机会

Consider it an opportunity

数到10，再问一次

Count to 10, ask again

A

以 后 再 处 理

Deal with it later

A

那 将 是 极 好 的

It's gonna be great

别有顾虑

Don't be concerned

追查一下其他线索

Follow something else's lead

A

你必须在
离开的时候处理好

You'll have to make it up as you go

找一个更清晰的视角

Get a clearer view

A

付出你的全部

Give it all you've got

A

要做就尽可能地
去做好，
不然就干脆别做

If it's done well; if not,
don't do it at all

A

未知可否

It is uncertain

A

不要忽视一般事实

Don't ignore the obvious

A

不要做得太过

Don't overdo it

别等了

Don't wait

A

享受这段经历

Enjoy the experience

如果你能
按照叮嘱的去做

If you do as you're told

A

其实有太多选择
跟几乎没有选择一样
让人头疼

Realize that too many choices
is as difficult as too few

A

不 会 失 败

It cannot fail

A

保持开明

Keep an open mind

必然的

It is certain

A

节省你的精力

Save your energy

这 并 不 重 要

It is not significant

A

告 诉 别 人 这 对 你 来 说 意 味 着 什 么

Tell someone what it means to you

当局者迷

You are too close to see

A

这很重要

It is significant

A

下不为例

Only do it once

A

这必然会使事情
变得更加有趣

It is sure to make
things interesting

A

你会获得最后的答复

You'll get the final word

好 事 多 磨

It is worth the trouble

A

舍弃老方案

Leave behind old solutions

A

此 事 可 能 已 成 定 局

It may already be a done deal

A

那将是浪费钱的

That would be a waste of money

A

那可能不容易，
但你会发现它的价值

It may be difficult
but you will find value in it

A

这将会影响
其他人对你的看法

It will affect how others see you

A

是的

Yes

A

它会带来好运的

It will bring good luck

A

这将会引起轰动

It will create a stir

A

负 责

Take charge

A

你最好把精力
放在自己的事情上

It would be better to
focus on the work

A

你将被消耗

It'll cost you

这正是
制订计划的好时候

It's good time to make plans

A

以轻松的步伐前进

Proceed at a more relaxed pace

A

你需要掌握主动权

You'll need to take the initiative

A

你该出发了

It's time for you to go

A

不 要 告 诉 别 人

Keep it to yourself

A

事态瞬息万变

The circumstances
will change very quickly

A

随 它 去 吧

Let it go

A

你 可 能 会 抵 触

You may have opposition

A

从不

Never

A

现在你可以

Now you can

当然

Of course

A

注意细节

Pay attention to the details

A

你会后悔的

You'll regret it

A

也许吧，当你成熟时

Perhaps, when you are older

可能吧

Maybe

A

不

No

A

以防万一

Prepare for the unexpected

A

你需要适应

You will need to accommodate

A

相关问题
将会浮出水面

Related issues may surface

A

你会发现
你所需要知道的一切

You will find out everything
you'll need to know

A

消除自己的障碍

Remove your own obstacles

遵守规则

Respect the rules

A

确定优先顺序
是过程的必要部分

Setting priorities will be a necessary part of the process

立即解决

Settle it soon

A

转移你的注意力

Shift your focus

A

结果可能是意料之外的

Startling events may occur as a result

把握机会

Take a chance

A

你必须妥协

You'll have to compromise

A

此事在你的控制之外

That's out of your control

A

答案可能
来自另一种语言

The answer may come to you
in another language

最佳方案
末必容易被理解

The best solution may not be
the obvious one

A

机 会 失 不 再 来

The chance

will not come again soon

A

结局将会是正面的

The outcome will be positive

A

保证不了

There is no guarantee

A

将会有困难
需要被克服

There will be obstacles to overcome

A

将会受到阻碍

There will be obstacles

A

这正是
制订计划的好时候

It's good time to make plans

无论你做什么，
结果都将是不可逆转的

Whatever you do,the results will be lasting

是的，但别勉强

Yes, but don't force it

你比以前更清楚了

You know better now
than ever before

你必须

You must

A

你会发现
自己无法妥协

You could find yourself
unable to compromise

把精力放在家庭生活上

Focus on your home life

A

调查了解之后
投入其中

Investigate and then enjoy it

那将仍是不可预测的

It will remain unpredictable

A

拭目以待

Watch and see what happens

A

这将会影响
其他人对你的看法

It will affect how others see you

A

此事在你的控制之外

That's out of your control

谨 慎 处 理

Approach cautiously

A

这将会影响
其他人对你的看法

It will affect how others see you

A

那将仍是不可预测的

It will remain unpredictable

绝对地

Definitely

A

如果你不抵触的话

If you don't resist

A

认 真 倾 听
你 就 会 知 道

Listen more carefully
then you will know

A

如果你能说“谢谢”

Provided you say "Thank you"

A

意 料 之 外

Startling events

A

那将仍是不可预测的

It will remain unpredictable

A

其他事物

将取决于你的选择

Others will depend on your choices

A

与另一种状况
息息相关

There is a substantial link to another situation

A

尽 你 所 能 去 进 步

Upgrade any way you can

A

你一定会得到支持

You are sure to have support

局势尚未清晰

The situation is unclear

A

找 出 更 多 办 法 来

Seek out more options

行 动 起 来

Move on

A

貌似是有把握的

It seems assured

A

不要浪费你的时间

Don't waste your time

A

耐 心 一 点

Be patient

A

大方一点

Be more generous

A

先去完成其他事

Finish something else first

A

那将是意外的

It could be extraordinary

那将是一件乐事

It will be a pleasure

A

此事不可取

It would be inadvisable

A

一 人 难 成 事

Not if you're alone

A

三思而后行

Take more time to decide

A

你需要考虑
其他办法
You'll need to consider other ways

你将得到支持

It will sustain you

A

不要被情绪左右

Don't get caught up
in your emotions

A

你会发现
自己无法妥协

You could find yourself
unable to compromise

A

尝 试 剑 走 偏 锋

Try a more unlikely solution

为确保做出
最好的决定，
请冷静

To ensure the best decision, be calm

相信你最初的想法

Trust your original thought

等一个更好的答复

Wait for a better offer

需要付出
足够的努力

A substantial effort
will be required

对 此 表 示 怀 疑

Doubt it

A

一年后这件事
就没那么重要了

A year from now it won't matter

把它写下来

Get it in writing

接受一个
日常上的改变

Accept a change to your routine

鼓起勇气

Adopt an adventurous attitude

A

把精力放在家庭生活上

Focus on your home life

A

最 好 再 等 等

Better to wait

A

你必须立即行动

You must act now

A

欣喜地确认这一点

Be delightfully sure of it

A

表现得就好像
这已经是真的了

Act as though it is already real

A

问问你爸

Ask your father

A

辅助将会让你的项目获得成功

Assistance would make your progress a success

尽早去做

Do it early

规避第一种方案

Avoid the first solution

A

把这当作一次机会

Consider it an opportunity

A

以后再处理

Deal with it later

A

别有顾虑

Don't be concerned

你必须在
离开的时候处理好

You'll have to make it up as you go

A

付出你的全部

Give it all you've got

A

未知可否

It is uncertain

不 要 做 得 太 过

Don't overdo it

A

享受这段经历

Enjoy the experience

A

其实有太多选择
跟几乎没有选择一样
让人头疼

Realize that too many choices
is as difficult as too few

保 持 开 明

Keep an open mind

A

节省你的精力

Save your energy

告诉别人这对你来说意味着什么

Tell someone what it means to you

这很重要

It is significant

A

这必然会使事情
变得更加有趣

It is sure to make
things interesting

A

好事多磨

It is worth the trouble

A

此事可能已成定局

It may already be a done deal

那 可 能 不 容 易 ，

但 你 会 发 现 它 的 价 值

It may be difficult

but you will find value in it

是的

Yes

A

这将会引起轰动

It will create a stir

你最好把精力
放在自己的事情上

It would be better to
focus on the work

A

这正是
制订计划的好时候

It's good time to make plans

你需要掌握主动权

You'll need to take the initiative

图书在版编目（CIP）数据

神秘的答案之书 / 韦恩著 . — 南京 : 江苏凤凰文艺出版社 , 2019.5 （2025.1 重印）
ISBN 978-7-5594-3522-4

Ⅰ . ①神… Ⅱ . ①韦… Ⅲ . ①人生哲学 – 通俗读物
Ⅳ . ① B821-49

中国版本图书馆 CIP 数据核字 (2019) 第 059457 号

神秘的答案之书

韦恩 著

责任编辑　刘洲原
特约编辑　王　蕾　申惠妍
装帧设计　韩　建
责任印制　刘　巍
出版发行　江苏凤凰文艺出版社
　　　　　南京市中央路 165 号，邮编：210009
网　　址　http://www.jswenyi.com
印　　刷　天津旭丰源印刷有限公司
开　　本　880 × 1230 毫米 1/32
印　　张　16
字　　数　10 千字
版　　次　2019 年 5 月第 1 版
印　　次　2025 年 1 月第 3 次印刷
书　　号　ISBN 978-7-5594-3522-4
定　　价　38.00 元